Physicalism & Feelings

An exploration of the nature of the universe

Jonathan Bloom

123 Books

*A catalogue record for this book is available
from the British Library*

ISBN: 978-1-907962-03-5

Published by 123 Books

Reading, England

For Galen

Contents

Preface

Most of us go about our day-to-day lives without reflecting too much on the nature of the universe. However, there seems to be a fairly widespread implicit assumption that the majority of the universe is very 'un-human like'. This assumption is often revealed when people talk about the majority of the universe being composed of 'matter' or 'physical stuff' or 'mere matter'.

It is assumed that humans are *not* 'mere matter' but that the universe which evolved humans and other life-forms *is* mere matter. So, if one tells people that one believes that there may be states in stones and tables which are analogous to the states in a human when that human claims to contain states of 'pain', then it is unlikely that other people will sincerely agree.

Why is this? What is this 'physical stuff' meant to be? And, why should one assume that, whatever it is, that it is most 'un-human like' when it is located in stones and tables? This book is an attempt to make some progress in answering these questions.

Introduction

Can a complete description of the universe be given in physicalist terms? The existence of subjectivity, of qualitative feeling, is often asserted to make this description impossible. To deal with feelings the physicalist either denies their existence, or asserts that they must somehow arise in the brain. The first approach is obviously wrong, whilst the second is explanatorily vacuous.

However, there is an alternative. The physicalist can account for the existence of qualitative feeling by asserting that all of the physical has feeling and that every physical interaction is a feeling interaction. In the chapters which follow I make the case for this alternative approach and I suggest that it is the best option for the physicalist to adopt.

Chapter 1

'Physical Stuff' & Knowledge

I assume that the word 'universe' refers to something that exists, and use the term to refer to the *aggregate of everything* that exists. I take 'physicalism' to be the view that the only type of stuff that exists in the 'universe' is 'physical stuff'. As part of the 'universe' can be described as 'mental phenomena' this means that, for the physicalist, 'mental phenomena' are 'physical stuff'.

How do we gain knowledge about this 'physical stuff'? And, how much can we know about the nature of this stuff – the 'physical'? We gain knowledge of the physical in a number of ways. Firstly, we are ourselves self-aware 'physical stuff' that through our awareness has direct knowledge of the nature of the physical. Secondly, we have sensory organs which give us mediated knowledge of the physical. Thirdly, we have scientific knowledge of the nature of the physical. These three knowledge generating mechanisms all provide a particular perspective on the nature of the physical. Together they can give us a good insight into the nature of the physical, but they are all individually limited perspectives. There seems to be no reason to believe that they can provide a complete description of all the 'physical stuff' in the 'universe'.

Our first mechanism of knowledge acquisition provides us with the insight that at least some 'physical stuff' has a quality that we can refer to as 'subjectivity', 'what it's likeness', or 'feeling'. I will refer to this quality of the physical as the quality of feeling. This mechanism also provides us with the insight that at least some of the physical has a quality that we can refer to as 'thought'. Our second mechanism of knowledge acquisition provides us with the insight that the physical has a quality that we can refer to as 'space-time occupation'. Our third mechanism of knowledge acquisition provides us with a much deeper view into the nature of the physical through the mathematical and technological tools of science.

Can science provide a complete description of the nature of 'physical stuff'? This possibility entails science giving a complete description of feeling. There are three different views as to whether this is possible. Firstly, it can be argued that future science will be sufficiently advanced that a description of the physical will be possible that includes *both* an 'activity in space-time occupation' element *and* a feeling element. In other words, it would be possible to describe a particular 'space-time occupation event' in terms of the feeling that was pertaining in that particular event. Secondly, it can be argued that feeling is completely describable in terms that make no reference to feeling. This means that a complete description of the physical would be possible in terms of 'activity in space-time occupation' alone. Thirdly, it can be argued that it is impossible, in principle, to give a complete description of feeling. In other words, feelings can only be known in their entirety by being felt, so science cannot give a complete description of the physical.

What are we to make of these three contrasting views? The second view is obviously wrong. However, both the first view and the third view have their merits. In support of the first view, it is surely possible that in the future neuroscience could advance to the point where feelings are perfectly correlated with particular 'space-time occupation events' in an 'aware subject'. However, this correlation requires the participation of the 'aware subject' who reports their feelings. So, if feeling is a more pervasive attribute of 'physical stuff' than 'aware subjects' this means that reportability will be impossible for some parts of the physical. It will be impossible to give a description of some 'space-time occupation events' in terms of the feeling that was pertaining in those events. In other words, the third view would be correct; it would be impossible, in principle, to give a complete description of feeling.

So, the answer to the question of whether science can, in principle, provide a complete description of the nature of the physical is dependent on the question of whether feeling is a more pervasive attribute of the physical than 'aware subjects'. If we assume that all that exists in the 'universe' is one type of 'physical stuff' then it has to be possible that feeling is more pervasive than 'aware subjects'. But how would we know if this was the case? And, if it is the case, how widespread could feeling be? Could all the 'physical stuff' of the 'universe' have a feeling aspect? This has to be a possibility. But again, how could we possibly know if this was the case?

If one were to hold that only some of the physical has feeling then one is carving up the 'universe' in a way that requires giving a satisfactory explanation as to *why only this part* of the physical has feeling. A naturalistic explanation of feeling requires such an explanation, but the formulation of such an explanation turns out to be highly problematic. For this reason I will defending the coherence of the alternative possibility that all of the physical has feeling; and that every physical interaction is a feeling interaction. I will be 'building up a picture' over the next six chapters, and hope that by the time you reach the conclusion that you will see the appeal of this alternative possibility.

Chapter 2

The 'Certainty' of Descartes

It is helpful to start by trying to understand why people might find the assertion that *every physical interaction is a feeling interaction* to be highly counterintuitive. A useful starting point is, perhaps, Descartes. In his quest for certainty Descartes argued that we can be certain that feelings exist. Furthermore, he argued that both thoughts *and* feelings are part of the human 'mind', which needs to be contrasted with the opposing 'physical' world which includes the human 'body'. The assertion that feelings exist *is* surely a certainty, but the assertion that feelings should be lumped together with thinking as unique aspects of a 'human mind' and opposed to the 'physical' is highly questionable. It is not certain that the feelings that exist are not simply aspects of the physical.

It is fashionable today to believe that the 'mind' is part of the natural order and should therefore be seen as physical rather than being opposed to the physical. Therefore, feelings also have to be seen as physical. However, despite this, the legacy of Descartes remains as physicalists still divide the world up into a *thinking/feeling aspect (or just 'thinking'!)* which they limit to the brain, and a *non-thinking/non-feeling aspect* which constitutes everything else. For much of human history prior to

Descartes it was assumed that feelings were located in the physical body; feelings were linked with the 'physical' rather than with thinking.

There are two types of physicalist today. One type *denies* the Cartesian 'certainty' by attempting to give a complete description of the 'universe' that makes no reference to feelings. Whilst, perhaps paradoxically, the other type acknowledges the reality of 'feelings' but *accepts* the Cartesian division by dividing the physical into a *thinking/feeling aspect* and a *non-thinking/non-feeling aspect*. Let us call the view that feelings exist but are limited to the brain 'braincentricism'.

Chapter 3

Anthropocentricism versus anthropomorphism

Feelings are uniquely human; the rest of the world is filled with automata, beasts and mere matter.

Is it not obvious that just as a human feels, contemplates and reasons, so does the cloud up in the sky above.

A human being is a truly tiny part of the universe. We observe the world around us and try and work out how it relates to us: How do I fit in? Which attributes do I share with the world that surrounds me? The answers to these questions change over time. When we look into the past we can see that there have been periods when the dominant outlook was what we would today describe as anthropomor-phic: humans were 'fallaciously' ascribing attributes of themselves to non-human 'physical stuff'. In contrast, we would view other periods as anthropo-centric: in these periods the dominant outlook 'fallaciously' denied the ascription of human attrib-utes to non-human 'physical stuff'. This means that human history can be viewed from the perspective of fluctuations in the boundary between anthropocen-tricism and anthropomorphism.

When one ponders recent realizations that have dethroned the specialness of man – Copernican cosmology, Darwinian evolution, and the Freudian unconscious – one would expect that the boundary would have moved away from anthropocentricism. However, perhaps paradoxically, these events which show man to be simply a part of the natural order, have led not to a greater appreciation of this fact, but rather to an increasing estrangement of man from his surrounding world. Perhaps the cause of this is that the 'dethroning events' have coincided with increasing human success in physics which has led to the belief that the outer world is predictable and the human world is pervaded by freedom.

The important point to note is that over time human perceptions about which of their attributes are uniquely human, and which are shared by the rest of the universe changes, and sometimes changes radically. In the pre-scientific age it was common to believe that the entire universe was thinking and feeling; whilst at the dawn of the 'scientific' age it was common to believe that feeling is a unique human attribute. In contrast, today feeling is commonly ascribed to many non-human animals but not to other parts of the physical. Our ascription of feeling to other parts of the physical seems to be based on the perceived similarity of those parts to ourselves. This is perhaps understandable, but it is surely a woefully inadequate way to ascribe feeling.

Is the braincentricism of feeling advocated by the physicalist anthropocentric? Or, is the assertion that every physical process is a feeling process anthropomorphising? There are no easy answers to these questions. However, what is certain is that the boundary between the anthropocentric and the anthropomorphic will change in the future. And a number of factors suggest that the shift will be away from anthropocentricism and possibly towards the ascription of feeling to every physical event. These factors include the growing contemplation of the previous 'dethroning events', advances in neurophysiology, and the recognition of the implications of both quantum mechanics and the Big Bang. This boundary shift wouldn't be a return to the pre-scientific age in which feeling and thinking were both ascribed to the universe as a whole. Rather, it entails a new division which supplants that of Descartes. In this new division feeling is ascribed to all 'physical stuff', whilst thinking is ascribed to a very small segment of 'physical stuff'.

Chapter 4

The Torch & the Brain

There is a widespread belief that there is something special about the brain that enables it to generate feeling. The existence of sensory organs seems to be one of the main causes of this belief. A sensory event is typically conceived of as a distinct feeling event; this conception can easily lead to the conclusion that the 'physical stuff' which lacks a brain also lacks feeling. For example, it is widely held that viewing a colour is a distinct feeling event, which has an associated state of "what it's likeness". These sensory feeling states of "what it's likeness" are referred to as qualia. If sensory organs did actually produce these distinct feeling states, if there was such a thing as a "pink quale", then it follows that the 'physical stuff' of the brain might have a capacity to generate feeling that other 'physical stuff' lacks.

However, the idea that there are distinct feeling states that arise from sensory organs in tandem with a brain could be utterly wrong. There could be no difference whatsoever between the feeling involved in the activity of a sensory organ and the feeling involved in the activity of other 'physical stuff'. Rather than there being a *distinctive feeling state* that is describable as a 'pink quale', there could be a plethora of 'feeling states arising from the interactions of 'physical stuff' *in tandem with* the creation of a wholly non-feeling appearance of 'pink'.

Let us consider the workings of two physical objects – a brain and a torch:

Colour creation in a brain

Patterns of light from the external world are focused on the retinas, which consist of three types of nerve-endings which are sensitive to long wavelengths of red light, intermediate wavelengths of green light, and short wavelengths of blue light. Electrochemical impulses from these nerve-endings are processed and travel to the visual cortex via the optic nerves.

Further processing in the visual cortex results in the creation of the colour 'pink'.

Light creation in a torch

The turning on of the switch from the external world causes the series circuit to close; this causes a current to travel from one battery terminal at the base of the torch to the other battery terminal which is located on the conducting metal body of the torch. The electrical current from the three batteries thereby reaches the cap of the lamp causing the filament of the bulb to glow. The parabolic-shaped reflector processes the radial rays of the bulb which results in the creation of 'parallel rays of light'.

On the face of it these two physical operations seem to be very similar. I will assume that it would be nonsensical to assert that there is a *quale of lightness* created when a torch creates light, but many philosophers do hold that there is a *quale of pinkness* created when a brain creates pink. If one wishes to assert that a *quale of pinkness* exists but that a *quale of lightness* doesn't, then one needs a very good reason; I don't believe such a reason exists. Rather, I wish to suggest that there is no sense whatsoever in which there is a feeling state of "what it's likeness" intrinsic to either the light creation of a torch or the pink creation of a brain.

Of course, there are physicalists who would wholeheartedly agree with this assertion; these are the physicalists who use the nonexistence of qualia to argue that feeling doesn't exist. However, this denies the 'certain' existence of feeling. There can be feeling states *without* sensory qualia. If every physical interaction is a feeling interaction this means that *every individual stage in the operation* of the brain and the torch is a distinct feeling interaction between physical entities. This means that the creation of pink and light are simply non-feeling side-effects of these underlying feeling interactions. These underlying feeling states will be in a constant state of flux whilst the pinkness and the light-creation stay constant. The feeling states that exist in a torch shouldn't be conflated with its ability to create light, and the feeling states in a brain shouldn't be conflated with its ability to create pink.

Chapter 5

The Objectiveness of Consciousness

Let us consider an objection that a qualia-believer might make to the comparison of a torch with a brain. It could be objected that there is a crucial difference because whilst the torch simply creates light, the creation of pink in a brain is accompanied by consciousness. In other words, 'light creation' is objective whilst 'pink' is actually a subjective creation of an aware subject. However, this objection is completely misguided. There is, in fact, nothing whatsoever 'subjective' about the creation of 'pink'.

'Pink' is simply a colour, an appearance, which is generated when a certain arrangement of 'physical stuff', such as a part of a human brain, interacts with another particular arrangement of 'physical stuff'. 'Pink creation' is a wholly objective, non-feeling, non-conscious event. It would be possible, in theory, to identify all of the 'physical stuff' involved in 'pink creation' and have complete knowledge that a particular part of the physical was currently creating 'pink'. If our scientific abilities reach this point then debates about the possibility of 'inverted qualia' / 'spectrum inversion' will be made redundant.

It is true that 'pink' creation can occur in parts of reality which have consciousness, such as a human brain, but consciousness itself is a transparent window, it doesn't itself endow what it becomes 'aware of' with subjectivity or feeling. This means that the lack of consciousness in a torch, the lack of a transparent window which engenders awareness of both the 'side-effects of physical arrangements' and the 'feelings in physical stuff', is irrelevant. There is no difference in kind between the physical processes of light creation in a torch and pink creation in a brain.

Chapter 6

The Labelling of Feelings

We have seen that a human possesses both the transparent window of consciousness and a complex language. The possession of these things can easily lead to a complete lack of comprehension as to how 'physical stuff' – such as a torch – can have feeling states. A reasonably intelligent person could assert: "I can feel love, feel excruciating pain, feel elation, and feel depressed. Only a fool would assert that a torch could have feelings like these."

So, let us level the playing field. Let us assume that all of the physical has feeling and that every physical interaction is a feeling interaction, we can then explore what the torch might say if it had both consciousness and a complex language. On being switched on the torch asserts: "The electrical current now running around my series circuit is excruciatingly painful, but when the filament of my bulb glows I explode into a state of elation, and when my parallel light rays illuminate my companion torch I feel a state of intense love."

This obviously sounds pretty ridiculous, but it highlights the utter ridiculousness of the position of those who assert that feeling states cannot be in a torch. Because human life is dominated by words we ascribe a word to some feelings we are aware of and call *that feeling state* say, 'elation'. Then we look at other 'physical stuff', say a 'torch', and ask: "Could a torch feel elation?" This is hardly a sensible question and is a hopeless way to address the issue of feeling. If a torch had consciousness and a language it would come up with words to describe its feeling states. Let us assert that one such feeling state would be called 'love'. The torch could then ask: "Could a human feel love?" I take it that a human would be incapable of feeling 'love' when the word refers to the feeling state of a torch. Just because a human has a battery of utterances at their disposal, and a torch doesn't, this doesn't mean that there aren't feeling states in a torch.

Chapter 7

The Fallacy of 'Objects'

I am contending that a physicalist could fruitfully believe that all 'physical stuff' has feeling and that every physical interaction is a feeling interaction. It needs to be stressed that this view entails that feelings are the domain of the fundamental constituents of physical reality; it would be erroneous to simply view feeling states as the attributes of arrangements of these constituents. The only objects that truly exist are the fundamental constituents of 'physical stuff', and it is interactions between these constituents that initiate changes in feeling states. It sounds a bit strange to assert that "there are feeling states in a torch" but what this actually means is "there are feeling states in the fundamental constituents of reality that constitute a torch".

It doesn't take very much serious reflection to realize that the only objects that exist are the fundamental constituents of physical reality. It requires a 'world creator' – such as a brain – to mould these constituents into larger 'objects' through its cognitive apparatus. To see this we only need to think of the Solar System and an atom. When we consider the Solar System we assert that the 'Moon' is an object, 'Mars' is an object, etc., we don't typically think of the Solar System as a whole as an individual object. However, when it comes to an atom we are happy to think of it as an individual object. Why is this? We can look up at the Moon and see it surrounded by space and thus we classify it as a distinct thing; we can also look through a techno-logical instrument see an atom and classify it as a distinct thing.

What are we to do when we realize that an atom is, in effect, a miniature solar system; that an atom is a tiny nucleus surrounded by circling electrons, and that both an atom and a solar system are 99 percent space? We have to conclude that it is our perspective on the world that creates 'objects' because the only difference between an atom and the Solar System is one of scale. If we were viewing from a higher perspective we would conclude that there are far fewer 'objects', and from a lower perspective we would conclude that there are immensely more 'objects'.

This realization enables one to see that the only objects that exist independent of a 'world creator' are the fundamental constituents of reality. When we are describing the feelings in 'physical stuff' there is no sense in which these feelings could be said to belong to anything other than the fundamental constituents. There are no 'feelings of a torch ', 'feelings of a stone', 'feelings of a dog' or 'feelings of a human'. It is simply the case that 'feelings' are in the fundamental constituents of 'physical stuff' and every physical interaction between this stuff is a feeling interaction.

Chapter 8

Conclusions

In attempting to give a complete description of the universe physicalists have either denied the 'certain' existence of feelings, or endorsed the explanatorily vacuous braincentricism. In this book I have explored an alternative way in which the physicalist can account for the existence of feeling states. The physicalist could fruitfully assert that all 'physical stuff' has feeling and that every physical interaction is a feeling interaction.

To support this position I have argued for a number of things. Firstly, there is no good reason why 'physical stuff' should be divided up into a *thinking/feeling aspect* and a *non-thinking/non-feeling aspect*. Rather, we could fruitfully assert that all 'physical stuff' is *feeling* and that only a very small part of 'physical stuff' thinks. Secondly, the boundary between the anthropomorphic and the anthropocentric changes and is likely to move away from the anthropocentric in the near future. Thirdly, the notion of feeling sensory qualia is erroneous; 'pink creation' is a wholly objective process; our feeling states need a different explanation. Fourthly, human consciousness and language make it difficult for reasonably intelligent people to believe/accept that all 'physical stuff' could have feeling states. And, finally, feelings are solely attributes of the fundamental constituents of 'physical stuff' because these are the only objects that truly exist.

This position doesn't entail that the whole universe is conscious or 'minded'. It simply entails prising apart the conceptual unity of thinking and feeling, asserting that feelings are an attribute of all 'physical stuff', and that thinking/mental phenomena/consciousness is a small part of 'physical stuff'. Furthermore, the position provides an elegant solution to the 'hard problem' of consciousness, as wherever the transparent window of consciousness exists it will illuminate feeling states, *because all 'physical stuff' has feeling states*. This seems to be a coherent position and to be the best way for the

physicalist to accommodate qualitative feeling.